Fran the Firefly has lost her light

An imprint by Firefly Island Publishers

For information about other storybooks and workbooks written by Serena Connor and related merchandise visit www.serenaconnor.com

Paperback ISBN 978-1-963921-18-2

Dear Parents

I trust that your child has enjoyed the captivating tale of Fran the Firefly in her quest to find her lost light. On this journey your child was introduced to new vocabulary and fascinating insights into pond habitats.

The "Fran the Firefly Student Workbook" accompanies the "Fran the Firefly has lost her light" storybook. This workbook seeks to reinforce and build on the concepts learned in the exciting tale of Fran. The subject areas covered in this workbook are wide ranging and include:

- Phonics
- Spelling
- Composition
- Reading Comprehension
- Literacy
- PSHE
- Science
- Puzzles and
- Colouring

The story of Fran is a fun and captivating tale that can be enjoyed by children of varying ages. The student workbook has been customized with this in mind. Each subject area comprises exercises with 3 levels of difficulty, catering to children with varying ability levels. We are confident that "Fran the Firefly has lost her light" is a book that your child will enjoy for years to come. As your child grows in competence, he or she will be able to tackle the higher level thinking exercises in this student workbook.

Exercises suitable for kids in Kindergarten and up.

Your support is invaluable, and I would be grateful if you could share your feedback by leaving a review on Amazon, Barnes and Noble, Apple and Kobo. Stay updated by liking and following me on YouTube, TikTok, Instagram and Facebook (Serena Connor the Author) and visit my website **www.serenaconnor.com** for more information about related merchandise and other storybooks and workbooks written by Serena Connor.

I eagerly anticipate sharing more adventures with you and your child. Your encouragement means the world to me.

Warm regards,
Serena Connor

First published in 2024 in the USA
Author: Serena Connor
Imprint: Firefly Island Publishers

About this workbook

The Fran the Firefly Student Workbook accompanies the "Fran the Firefly has lost her light" storybook. The workbook has exercises of varying levels of difficulty.

Level 1

Pages that have a yellow flag are of the lowest difficulty level. This level is suitable for beginner readers, ages 4 to 6.

Level 2

Pages that have a blue flag are of medium difficulty level. This level is suitable for more independent readers, ages 7 to 8.

Level 3

Pages that have a green flag are of the highest difficulty level. This level is suitable for more competent readers, ages 9 to 10.

As your child grows in competence, he or she will be able to tackle the Level 2 and 3 exercises in this book.

HAPPY LEARNING

NAME ______________________

GRADE ______________________

TIME TO LEARN

Keep on TRYING

learn SOMETHING new EVERYDAY

STAY CREATIVE!

Phonics

Be GRATEFUL everyday

PHONICS

Fill in the missing letters

SU__

SHEE__

NEW__

SNAI__

TRE__

PHONICS

Fill in the missing letters

__OUSE

__ EAF

__LOWERS

__LANE

__OAT

PHONICS

Fill in the missing letters with O or U

P___ND

FR___G

B___G

D___CK

C___W

PHONICS

Fill in the missing letters with A or E

W__T

FR__N

GR__SS

SUNS__T

UPS__T

PHONICS

Fill in the missing letters with I or E

P __ P

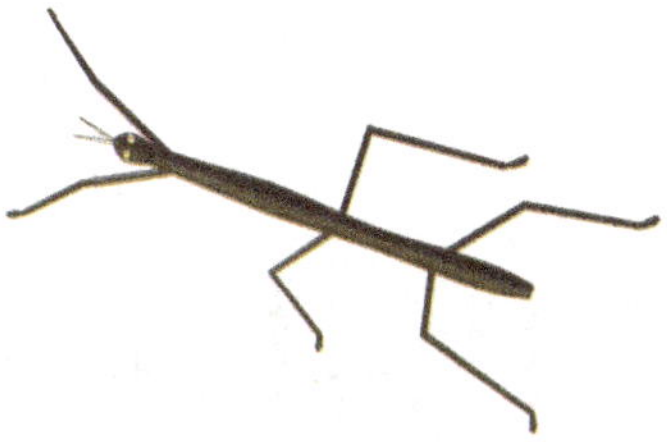

FR __ T

W __ NG

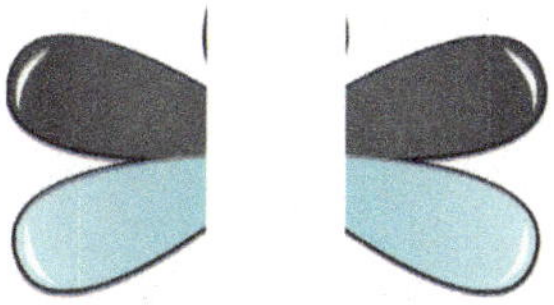

L __ LY PAD

D __ M

PHONICS

Can you call these words that end with

"ight"

light

night

tight

sight

fight

whight

fright

bright

KEEP IT UP!

PHONICS

Fill in the missing letters

__ IGHT

__ IGHT

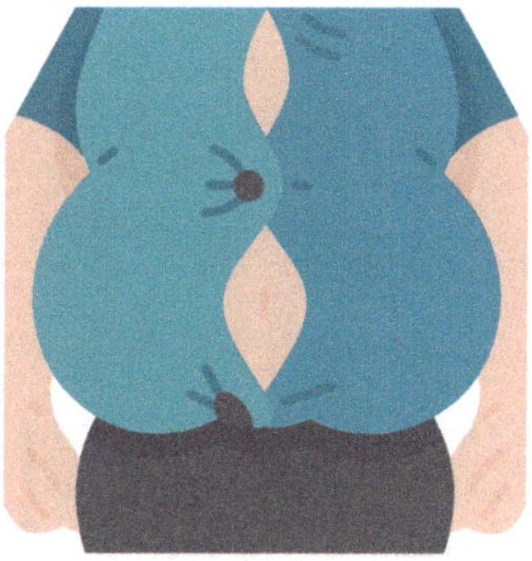

__IGHT

KEEP IT UP!

PHONICS

Fill in the missing letters

___IGHT

___ ___IGHT

___ ___IGHT

KEEP IT UP!

Spelling

FRAN THE FIREFLY
WORD SCRAMBLE

Unscramble the following words

OPDN ____________________

GOFR ____________________

YSK ____________________

ERTE ____________________

KUDC ____________________

GBU ____________________

KEEP IT UP!

FRAN THE FIREFLY
WORD SCRAMBLE

Unscramble the following words

CROK ____________________

EHOSU ____________________

TABO ____________________

TGHLI ____________________

LASIN ____________________

EHEPS ____________________

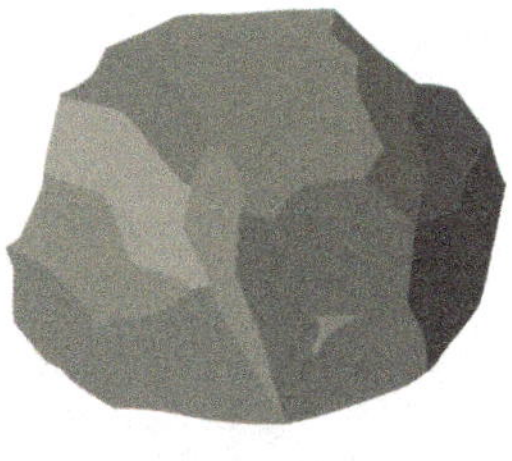

KEEP IT UP!

FRAN THE FIREFLY
SPELLING

Circle the correct spelling

PON POND PONDE

TREE THREE TREI

BOTE BOAT BAOT

ROK ROCK RUCK

LITE LIGTH LIGHT

HOWS HOUSE HOUS

FRAN THE FIREFLY
SPELLING

Circle the correct spelling

SNAL SNAIL SNALE

SHEAP SHEPE SHEEP

SKI SKYE SKY

BUG BOG BURG

FROG FROGE FROGG

DUCK DUK DOCK

KEEP IT UP!

FRAN THE FIREFLY

WORD SCRAMBLE

Unscramble the following words

(Look at the next exercise for clues)

NPDO KTARSE	P ____________
GOFR	F ____________
LFOSURNETCE	F ____________
ELGE	G ____________
ISLNA	S ____________
FFETDINRE	D ____________
TGIHLUNS	S ____________
ASEONSS	S ____________
RKELSAP	S ____________
WROFN	F ____________
TFER	F ____________
ZBIZGNU	B ____________

FRAN THE FIREFLY

ANSWER THE FOLLOWING QUESTIONS

PIP IS A ____________________

FREDDY IS A ____________________

DESCRIBES A BRIGHT LIGHT ____________________

ANOTHER WORD FOR HAPPINESS ____________________

SNIPPY IS A ____________________

OPPOSITE OF SAME ____________________

LIVING THINGS NEED THIS ____________________

SUMMER IS ONE OF THE ____________________

SHINE BRIGHTLY WITH FLASHES OF LIGHT ______________

FACE YOU MAKE WHEN UPSET____________________

TO BE VISIBLY ANXIOUS____________________

SOUND MADE BY BEES____________________

Composition

Composition

Use the words below to complete the story

Fran firefly pond light

lost help me

My name is ______ . I am a __________.

I live by the _________. I cannot see

my _______. I think that it is _______.

I need your ________. Will you please

help _____. Thank you very much.

Persuasive Writing Tips

Use these frames to make a plan for your persuasive writing.

HELPFUL TIPS

1. Who are you asking?
2. What do you want?
3. Why do you want it?
4. When did you last see your light?
5. How can they help?

Why do you need your light?

Usefulness (I need to see in the dark)

Aesthetic reasons (It makes me look beautiful)

Emotive reasons (I miss my light/ I feel incomplete)

Belonging (All fireflies have lights)

Useful descriptive words

colourful?
bright ?
mesmerizing ?
iridescent?
devastated?
lost?
sad?

Don't forget your manners! You are asking for help.

-Please can you...
-I would greatly appreciate it if you could...
-Thank you

Composition

Pretend that you are a firefly. Write a persuasive paragraph trying to convince the other pond animals to help you to find your light.

Don't forget to use your adjectives!!

Composition

Pretend that you are a firefly. Write a couple persuasive paragraphs trying to convince the other pond animals to help you to find your light.

Don't forget to use your adjectives!!

you can do it!

Don't forget to use your adjectives!!

you can do it!

REMEMBER that you are AMAZING

Reading Comprehension

Reading Comprehension

CIRCLE THE CORRECT ANSWER

1.Fran is a

2. Fran lives in a

3. Fran cannot find her

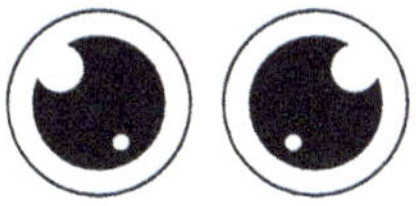 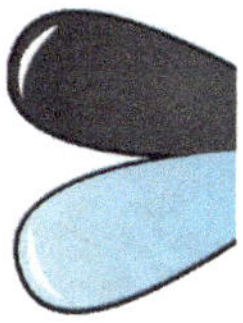

Remember that you are amazing

Reading Comprehension

CIRCLE THE CORRECT ANSWER

4. Who was eating flies?

5. Who sat on a big rock?

6. What colour is Nathan the Newt?

Reading Comprehension

CIRCLE THE CORRECT ANSWER

1. Fran lives in a sea/pond.

2 Fran's light disappeared because it was dark.
True or false

3. When it was very dark, Fran's light was
a) bright
b) dim

4. Mozzie is a
a) fly
b) mosquito
c) cricket

5. What was Fred the frog doing when Fran saw him?
a) eating
b) jumping
c) sleeping

6. Reeds are a type of
a) plant
b) animal

7. Fran found her light when it was bright outside.
True or False

8. When Fran found her light she felt
a) happy
b) disappointed

9. Fran is a dragonfly. True or False

10. Fran's story teaches us never to give up. True or False

great job

Reading Comprehension

CIRCLE THE CORRECT ANSWER

1. Fran lives in a ______.

2. Why did Fran's light disappear?
a) because she was tired
b) because it was no longer dark
c) because it was raining
d) because someone stole her light

3. When it was very dark
a) Fran's light was small and dim
b) Fran's light was not very bright
c) Fran's light was big and bright
d) Fran's light was dim and round

4. Which word best describes the mosquito?
a) angry
b) tired
c) quiet
d) happy

5. Fred the frog was
a) helpful
b) hungry
c) tired
d) lonely

6. Why was Fran exhausted?

a) because she was flying for a long time.

b) because something was wrong with her wings.

c) because her wings were not strong.

d) because she was ready to give up.

7. Reeds are tall leaved plants in the grass family. True or False

8. Why did Fran's light shine brightly at the end of the story?

a) it was now dark

b) it was raining

c) she was happy

d) she turned the switch on

9. Circle the WORDS that best describe how Fran felt when she found her light.

a) proud

b) relieved

c) angry

d) excited

10.The moral of the story is

a) don't ever be sad

b) never give up

c) you will always get help

d) give up when you are tired

you are INCREDIBLE

LITERACY

OPPOSITE

SYNONYM

ALLITERATION

ONOMATOPOEIA

DEFINITION

YOU GOT THIS! Let's refresh our memories

OPPOSITE

A word that has a different meaning to another word

E.G. MEAN - NICE
GIGANTIC-TINY
HOT-COLD

SYNONYM

A word that has the same or similar meaning as another word

E.G. HUGE - ENORMOUS
UNHAPPY- SAD
TIRED-EXHAUSTED

LANGUAGE

OPPOSITES

Match the OPPOSITES

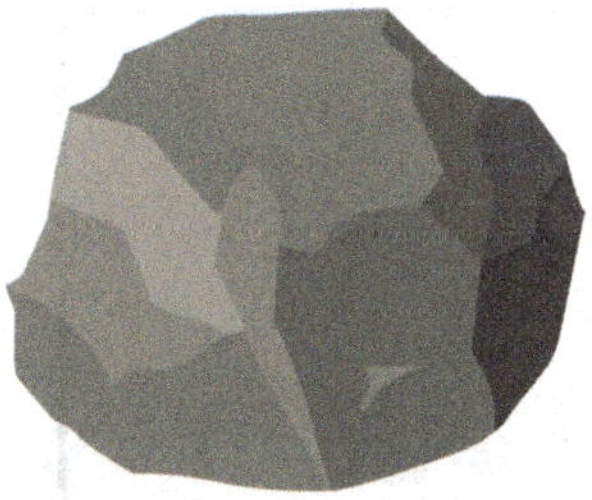

big

up

hot

happy

day

cold

small

night

sad

down

YOU GOT THIS!

LANGUAGE

SYNONYMS

Underline the synonyms to the words in purple

JUMP

hop sit

HAPPY

sad joyful

UNHAPPY

sad mad

ANGRY

sorry mad

HUGE

big round

SMALL

little round

YOU GOT THIS!

LANGUAGE

OPPOSITES

Underline the OPPOSITES to the words in the circles

DAY	NIGHT	BRIGHT
UP	SIDE	DOWN
DIM	BRIGHT	LIGHT
WET	DAMP	DRY
LOST	MISSING	FOUND
HOT	COLD	STEAMING

LANGUAGE

SYNONYMS

YOU GOT THIS!

Underline the SYNONYMS to the words in the circles

SAD	ANGRY	UPSET
BIG	ROUND	ENORMOUS
DIM	BRIGHT	DULL
WET	DAMP	DRY
LOST	MISSING	FAR
NEAR	CLOSE	BY

Let's refresh our memories

LANGUAGE

ALLITERATION	Repetition of the same consonant sound at the beginning of words	E.G. Fran the fussy firefly flew far away to Finland.
ONOMATOPOEIA	Onomatopoeia is a word that sounds like the noise that is being described	E.G. SPLAT, POW, BOOM, THUD
SYNONYM	A synonym is a word or phrase that has the same or similar meaning as another word or phrase	E.G. BIG - ENORMOUS UNHAPPY- SAD TIRED-EXHAUSTED
ANTONYM/ OPPOSITE	An antonym is a word of opposite meaning to another word	E.G. MEAN - NICE GIGANTIC-TINY HOT-COLD
DEFINITION	A definition is the meaning of a word or a term	E.G. RUN MEANS TO MOVE AT A SPEED FASTER THAN A WALK, NEVER HAVING BOTH OR ALL FEET ON THE GROUND AT THE SAME TIME.

Let's practice what we have learnt

Match

ALLITERATION	PLOP
ONOMATOPOEIA	ginormous - tiny
SYNONYM	Nosey Nathan Newt was naughty
ANTONYM/ OPPOSITE	A mosquito is a small flying insect
DEFINIITION	furious - angry

I DID + MY + BEST !

Let's practice what we have learnt

Match

ALLITERATION	worry- fret
ONOMATOPOEIA	bleak means without hope
SYNONYM	BUZZZZ
ANTONYM/ OPPOSITE	Happy Harry has a heavy hat
DEFINIITION	exhausted- energetic

Personal, Social and Health Education

1	2	3	4	5

very unhappy	a bit unhappy	okay neither happy nor sad	quite happy	very happy

Personal, Social, Health Education

Fran experiences many different emotions throughout this story. Listed below are some of the emotions experienced by Fran. In relation to each emotion, do you recall why she was feeling that particular way. What about you? When do you feel the emotions listed below. Discuss with your teacher or your parent.

shocked

fearless

anxious

determined

calm

sad

frustrated

tired

proud

Mood Check. On a scale of 1 to 6, how happy are you today? What can you do today that would make you feel very happy?

1	2	3	4	5
very unhappy	a bit unhappy	okay neither happy nor sad	quite happy	very happy

Personal, Social, Health Education

FEELINGS

Match the words to show how Fran is feeling

TIRED

SAD

HAPPY

ANGRY

Science

Science

Match to complete the sentence

Mozzy is a

frog 

Pip is a

pond skater 

Fred is a

mosquito 

Nathan is a

snail 

Snippy is a

newt 

great job

Science

Circle the correct description for the pictures below

plant or animal

plant or animal

plant or animal

plant or animal

plant or animal

plant or animal

plant or animal

plant or animal

Science

ANIMALS

Match the missing parts to the correct animal

Science

PLANTS

Circle the plants found in the story of Fran

coconut tree

reeds

flowers

cherry blossom tree

grass

christmas tree

lily-pads

bamboo tree

Science

ANIMALS

All birds and most insects have wings and can fly.

Can you circle all of the animals that can fly?

you are
A STAR

Science

Animals and plants are living things. All living things grow, reproduce and need nutrients and water to live.

Circle the correct description for the pictures below?

living or non-living

living or non-living

living or non-living

living or non-living

living or non-living

living or non-living

living or non-living

living or non-living

Science

PLANTS

1. Write the correct word under the pictures below

grass tree lily-pad
flowers algae reeds

Keep learning

Science

ANIMALS

Write the correct word under the pictures below

duck pond skater mosquito firefly
ladybug butterfly

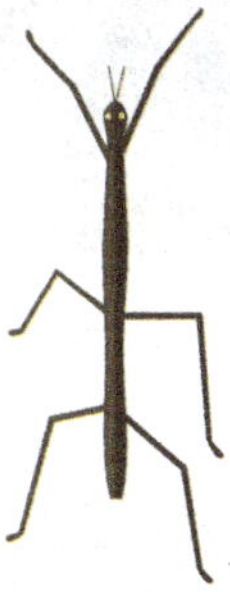

YES YOU CAN !!!

Science

LIGHT

Match the firefly to the appropriate picture

Science

PLANTS

Plants, like animals are living things. They grow from the soil and turn sunlight into food in their leaves. In addition to sunlight, plants need water, nutrients and carbon dioxide to make their food and to grow. They get water and nutrients from the soil through their roots and absorb carbon dioxide from the air through their leaves. Once energy is produced, plants release oxygen through their leaves. The process by which plants use sunlight, carbon dioxide and water to create energy is called photosynthesis.

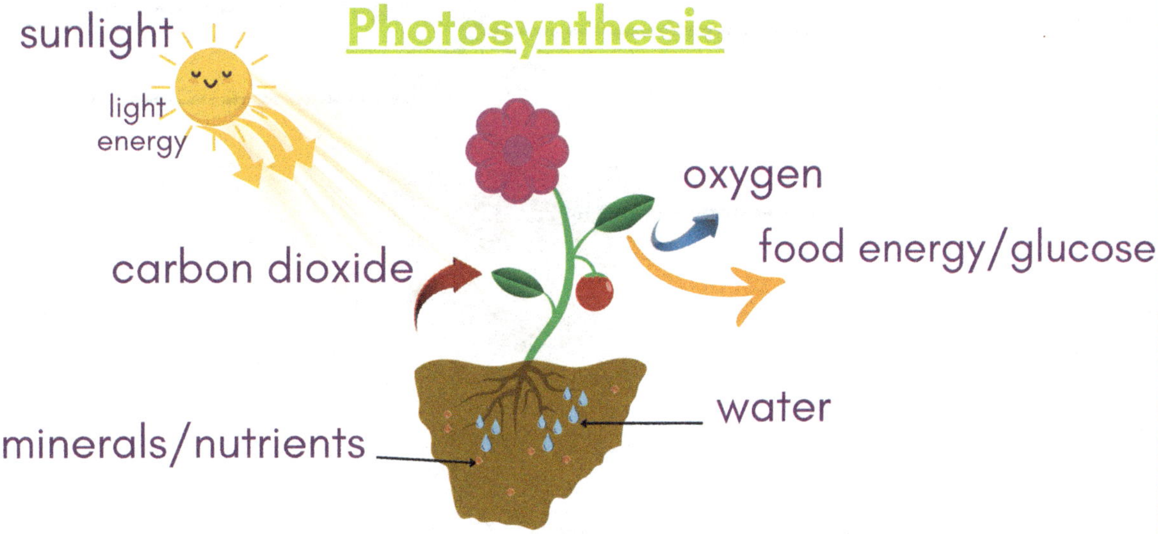

Did you know that algae are not plants?

Some algae, for example seaweed, look like plants. However, algae are actually neither plants nor animals. Algae belong to a group of living things called protists.

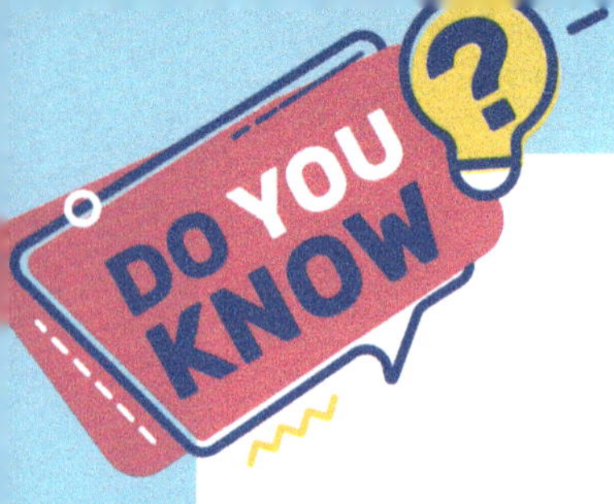

Science

PLANTS

Use the words in the box to label the plant and to answer the questions below

flower	leaf	air	soil
stem	fruit	roots	

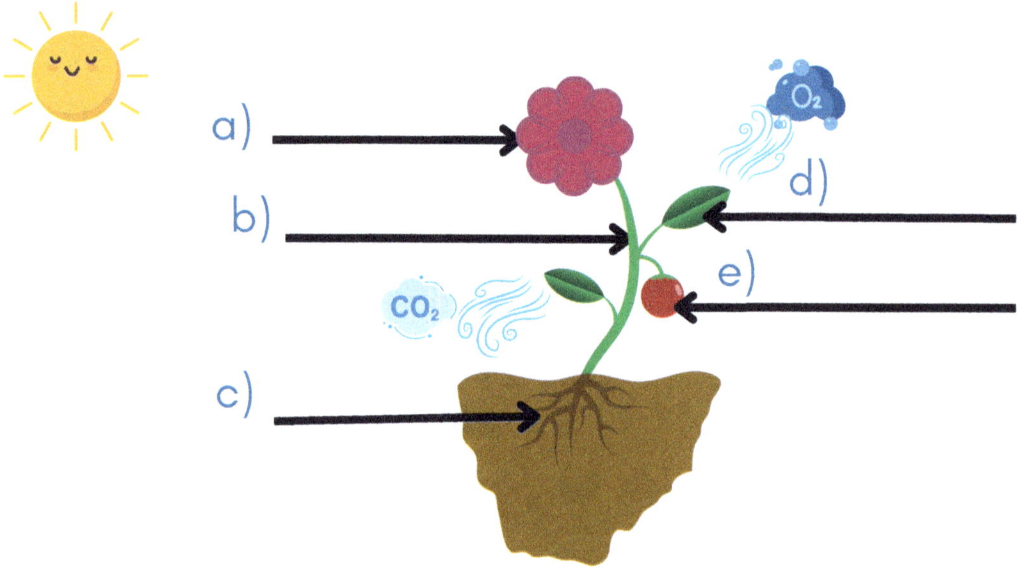

1 The _____ keeps the plant erect.

2. Plants get nutrients and water through their ______

3. Sunlight is turned into food in this part of the plant _____

4. The leaves absorb carbon dioxide from and release oxygen into the ___

5. Minerals and water are found in the _____

your hardwork pays off

Science

PLANTS

Write the names under the pictures below

Answer the questions below

1. Plants make their own food using water, sunlight and carbon dioxide. True or False
2. Which plant is a type of grass ?_____
3. Which plant is a home for squirrels, snakes and birds?____
4. All of the above are examples of flowering plants. True or False
5. This flowering plant floats on the surface of the pond. ____ ___
6. Seaweed is a type of algae. True or False
7. Algae is a type of plant. True or False
8. Plants get their nutrients from the sun. True or False

DO YOU KNOW

Science

ANIMALS

Animals are living things. Animals can be divided into five distinct groups: mammals, fish, birds, reptiles and amphibians.

Animals need food and water for energy. All living animals eat, drink, breath, excrete, move, grow and reproduce.

DO YOU KNOW

Science

ANIMALS

Match the animal to the appropriate category

Mammal

Fish

Amphibian

Reptile

Bird

eagle

stingray

turtle

frog

dolphin

NEVER GIVE UP

Science

ANIMALS

Write the names of the animals below

Answer the questions below

1. All of the above animals can fly. True or False
2. Which of the above animals is not an insect?____
3. None of these animals live in or by a pond? True or False
4. All of these animals reproduce. True or False
5. Which animal lights up when it is dark? ________.
6. All of the above animals lay eggs. True or False
7. All animals m___, b_______, e_______, g___, e__ and d____.

Science

FOOD CHAINS

THIS → IS EATEN BY → THIS

A food chain describes the order in which organisms and living things depend on each other for food. Every ecosystem or community of living things, has one or more food chains. Most food chains start with organisms that make their own food, such as plants.

DON'T BE AFRAID TO FAIL, BE AFRAID NOT TO TRY

Science

FOOD CHAINS

Match the consumer to the appropriate food

THIS → EATS → THIS

Science

HERBIVORE, CARNIVORE, OMINIVORE

HERBIVORE

A herbivore is an animal that only eats plants.

CARNIVORE

A carnivore is an animal that only eats other animals

OMINIVORE

An omnivore is an animal that eats both plants and animals

Science

Fill in the box with the correct description

HERBIVORE, CARNIVORE, OMINIVORE

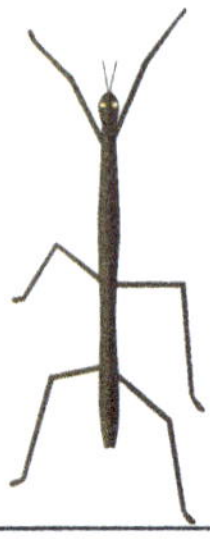

Keep on TRYING

Science

Name 4 plants found in the story of Fran the Firefly

1. g____
2. t____
3. r____
4. l______

Name 5 animals found in the story of Fran the Firefly

1. d___
2. f___
3. m_______
4. f______
5. s____

Name 5 non-living things found in the story of Fran the Firefly

1. b___
2. h____
3. k___
4. r___
5. c__

Name 4 insects found in the story of Fran the Firefly

1. m________
2. b_________
3. f_______
4. p__________

STAY FOCUSED!

Science

Facts about insects

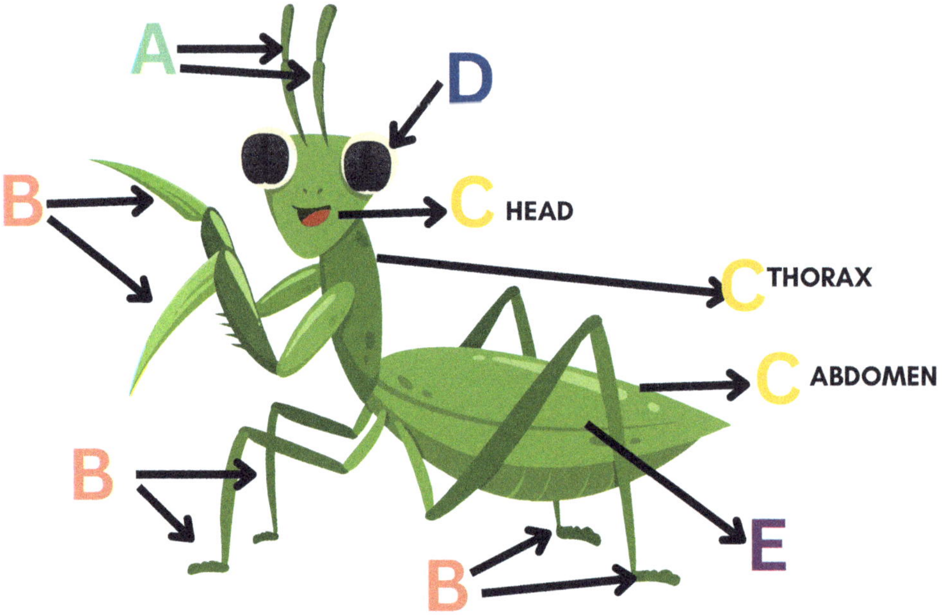

A) All insects have a pair of antennae.

B) All insects have 6 legs.

C) All insects have a 3 part body:
head, thorax and abdomen

D) All insects have compound eyes which consist of hundreds or thousands of tiny light-sensitive parts called ommatidia.

E) Most insects have wings.

STAY FOCUSED!

Science

Facts about insects

A) Some insects have a 2 part body. True or False

B) All insects have 8 legs. True or False

C) All insects have a pair of antennae. True or False

D) Most insects have wings. True or False

E) Circle the insects?

F) Which of the above animals are not insects?

1.
2.

G) How do you know that these animals are not insects?

H) Can you write the names of 5 insects?

1.
2.
3.
4.
5.

Puzzles

Fran the Firefly Word Search

Find the 6 hidden words. Words can be found across and down.

F	I	R	E	F	L	Y
U	P	H	I	E	R	B
L	S	T	S	N	A	U
P	K	R	H	T	B	G
O	Y	R	D	U	C	K
N	A	A	O	E	I	E
D	I	T	R	E	E	S

POND

BUG

FIREFLY

DUCK

SKY

TREE

Fran the Firefly

WORD SEARCH

Find the hidden words.

F	L	U	O	R	E	S	C	E	N	T	C
B	D	S	P	N	A	L	O	C	E	A	S
E	I	O	H	T	R	V	F	E	L	R	U
E	F	O	G	F	H	O	L	T	K	P	N
T	F	W	M	N	N	U	N	E	R	A	T
L	E	M	T	N	A	E	H	B	A	R	F
E	R	B	E	S	R	I	T	A	P	T	L
S	E	N	R	E	N	L	L	A	S	Y	I
D	N	O	P	S	K	A	T	E	R	O	T
K	T	M	O	S	Q	U	I	T	O	I	T
G	O	R	F	I	R	E	F	L	Y	F	E
B	P	R	E	S	E	N	T	I	R	E	D

DIFFERENT	SPARKLE	FROG	BEETLES
MOSQUITO	FIREFLY	FLITTED	SNAIL
SUN	POND	FLUORESCENT	TIRED

I'm PROUD

Fran the Firefly

CROSSWORD PUZZLE

duck   pond skater   mosquito   fly   butterfly   firefly

1. What type of insect is Pip?
2. What type of insect is Mozzy?
3. This insect has colourful wings.
4. This insect glows in the dark.
5. This animal has a beak.
6. Animals with wings can do this.

Fran the Firefly

CROSSWORD PUZZLE

1. This insect's name is Mozzy
2. Plants need this to grow
3. To move from place to place very quickly
4. Fred the frog was very ______
5. Fran the firefly lost her ______
6. Fran sat on the leaf because she was ______
7. Tears fell from Fran's _____
8. Freddy noticed that Fran was very ________
9. Opposite of day
10. A firefly is a type of ______

Maze

Help Fran find her light

Maze

Help Snippy find his way through the mazes

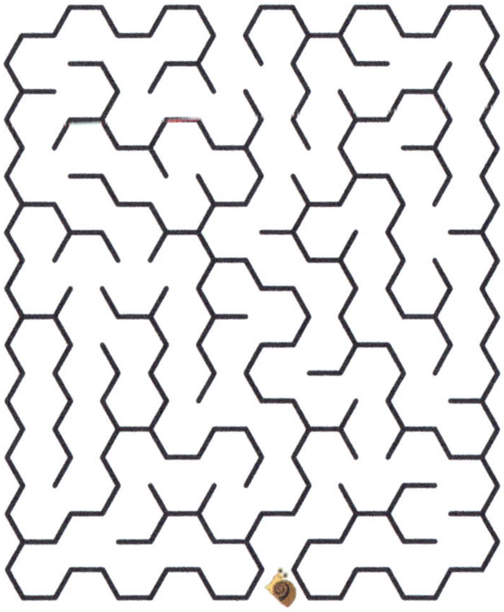

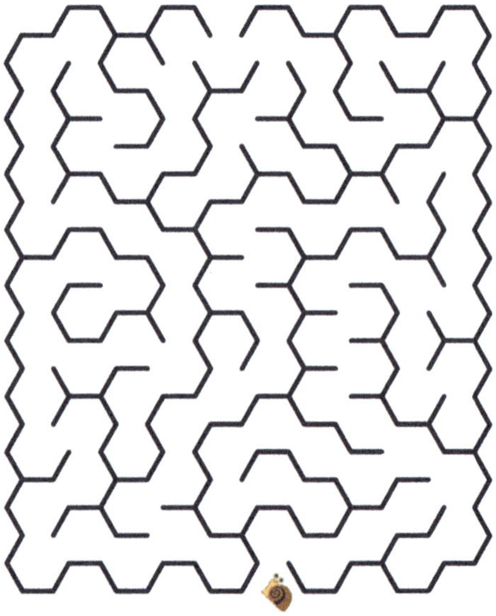

Maze

Help Mozzie find her way through the maze

Colour these insects

Have Fun

COLOURING-IN

Dear Parents

Pg 7
Sun
Sheep
Newt
Snail
Tree

Pg 8
House
Leaf
Flowers
Plane
Boat

Pg 9
Pond
Frog
Bug
Duck
Cow

Pg 10
Wet
Fran
Grass
Sunset
Upset

Pg 11
Pip
Fret
Wing
Lily-Pad
Dim

Pg 13/14
Light
Night
Tight
Sight
Bright
Fright

Pg 16
Pond
Frog
Sky
Tree
Duck
Bug

Pg 17
Rock
House
Boat
Light
Snail
Sheep

Pg 18
Pond
Tree
Boat
Rock
Light
House

Pg 19
Snail
Sheep
Sky
Bug
Frog
Duck

Pg 20 (same for 21)
Pond skater
Frog
Fluorescent
Glee
Snail
Different
Sunlight
Seasons
Sparkle
Frown
Fret
Buzzing

Pg 23
Fran
firefly
pond
light
lost
help
me

Pg 29/30
1) 
2) 
3)
4) 
5) 
6) 

Pg 31
1) pond
2) false
3) a)
4) b)
5) a)
6) a)

Pg 32
7) False
8) a)
9) False
10) True

Pg 33
1) pond
2) b)
3) c)
4) d)
5) b)

Pg 34
6) a)
7) true
8) a)
9) a) b) d)
10) b)

Pg 37
big-small
up-down
hot- cold
happy- sad
day-night

Pg 38
jump-hop
happy-joyful
unhappy-sad
angry-mad
huge-big
small-little

Pg 39
day-night
up-down
dim-bright
wet-dry
lost-found
hot-cold

Pg 40
sad- upset
big-enormous
dim-dull
wet-damp
lost-missing
near-close

Pg 42
Alliteration- Nosey Nathan Newt was naughty
Onomatopoeia- PLOP
Synonym - furious-angry
Antonym/opposite- ginormous-tiny
Definition- A mosquito is a small flying insect

Dear Parents

Pg 43

Alliteration- Happy Harry has a hat
Onomatopoeia- Buzzz
Synonym- worry-fret
Antonym- exhausted-energetic
Definition-bleak means without hope

Pg 46

Tired
Sad
Happy
Angry

Pg 48

Mozzy is a- mosquito
Pip is a - pond skater
Fred is a -frog
Nathan is a - newt
Snippy is a - snail

Pg 49

- animal
- animal
- plant
- animal
- plant
- plant
- animal
- animal

Pg 50

Pg 51

- reeds
- flowers
- grass
- lily-pads

Pg 52

Pg 53

- non-living
- living
- living
- non-living
- living
- living
- living
- non-living

Pg 54

- grass
- reeds
- flowers
- tree
- lily-pads

Pg 55

- firefly
- butterfly
- ladybug
- pond skater
- mosquito
- duck

Pg 56

Pg 58

a) flower	1. stem
b) stem	2. roots
c) roots	3. leaf
d) leaf	4. air
e) fruit	5. soil

Pg 59

- algae
- reeds
- flowers
- tree
- grass
- lily-pad

1. true
2. reeds
3. tree
4. false
5. lily-pad
6. true
7. false
8. false (soil)

Pg 61

mammal- dolphin
fish- stingray
amphibian- frog
reptile- turtle
bird- eagle

Dear Parents

Pg 62

firefly
butterfly
ladybug
pond skater
mosquito
duck

1. true
2. duck
3. False
4. True
5. firefly
6. true
7. move, breathe, excrete, grow, eat and drink

Pg 64

Pg 66

mosquito - omnivore
pond skater- carnivore
sheep- herbivore
butterfly-herbivore
frog- carnivore
snail-omnivore
rabbit- herbivore

Pg 67

1. grass
2. trees
3. reeds
4. lily-pads

1. duck
2. frog
3. mosquito
4. firefly
5. snail

1. boat
2. house
3. kite
4. rock
5. car

1. mosquito
2. butterfly
3. firefly
4. pond skater

Pg 69

A) False
B) False
C) True
D) True
E) 
F) scorpion and spider
G) They do not have 6 legs
H) mosquito
ladybug
firefly
roach
bee
butterfly
fly
grasshopper
dragonfly
pond skater

Pg 73

1. pond skater
2. mosquito
3. **butterfly**
4. firefly
5. duck
6. fly

Pg 74

1. mosquito
2. sunlight
3. **flit**
4. hungry
5. light
6. tired
7. eyes
8. worried
9. night
10. beetle

Serena Connor is a passionate professional with a background in Education and Law, hailing from the picturesque island of Anguilla in the British West Indies. A former Kindergarten teacher and a mother of two spirited girls, Serena has a passion for storytelling and education. Her love for bedtime reading with her daughters inspired her to write 6 captivating educational children's books that comprise the "Lost and Found Collection." This collection kicks off with "Fran the Firefly Has Lost Her Light", a delightful tale filled with beautiful characters and engaging rhymes that children and parents can enjoy together, time and time again. A workbook that comprises exercises from varying subject areas, accompanies each of the storybooks in this collection. In addition to her literary pursuits, Serena is a dedicated lawyer at the Attorney General's Chambers, where she drafts legislation which governs the island. Furthermore, she is the proud owner of Good Korma, a renowned Pan-Asian restaurant on the island, where she shares her love for diverse Asian cuisines with the community.

For more insights into Serena Connor's creative works and updates on storybooks, workbooks and related merchandise in the Lost and Found Collection, you can follow her on YouTube, Facebook, TikTok and Instagram under the handle SerenaConnortheAuthor or visit www.serenaconnor.com. Serena's dedication to the legal profession and the culinary world, combined with her passion for children's literature, showcases her diverse talents and unwavering commitment to enriching the lives of those around her.

OTHER BOOKS IN THE LOST AND FOUND COLLECTION

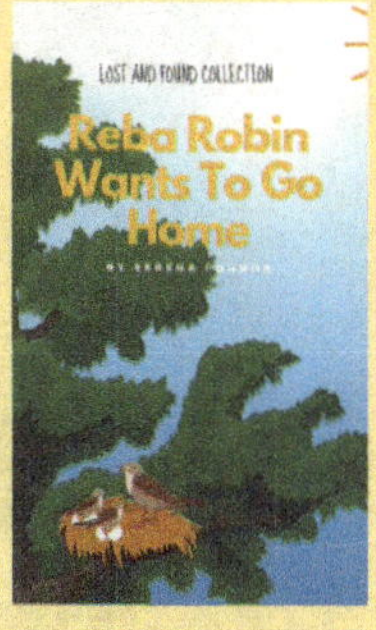

TEACHER MALICA SMITH-CONNOR

Tr. Malica
Remediial Reading
teacher
Anguilla

The Fran the firefly workbook is designed to assist with the development of critical thinking skills and can be used by the child as he or she advances academically, with exercises ranging from low, medium to high difficulty levels. This book focuses on various subject areas, which include Phonics, Spelling, Composition, Literacy and Science and can be used in school and at home to support learning in the aforementioned subject areas. I look forward to doing the Level 1 exercises with my remedial reading students.

PRINCIPAL ELECTRA BUDDLE

"Fran the Firefly Has Lost Her Light" is a delightful and educational children's book that combines storytelling with science. The use of rhyme and repetition throughout the story not only makes the tale enjoyable to read aloud but also reinforces memory for early learners, helping them retain the scientific term and story details. This story is a wonderful addition to any child's library.

Electra Buddle
Principal of the
Alwyn Allison
Primary School
Anguilla

READING SPECIALIST, CHRISSY GATES

Chrissy Gates
Reading Specialist
Washington D.C.

Fran the Firefly is a beautiful story teaching children that even when they feel darkness, they have a beautiful light to shine. This storybook uses rhyme to share the story of Fran, seeking her light. The accompanying workbook incorporates the Science of Reading research and creates phonological activities combined with reading comprehension skills to reinforce early reading. An excellent book to share with our youth!

LOOK OUT FOR THE ACCOMPANYING WORKBOOKS IN THE LOST AND FOUND COLLECTION